YOUR KNOWLEDGE HAS VALUE

- We will publish your bachelor's and
  master's thesis, essays and papers

- Your own eBook and book -
  sold worldwide in all relevant shops

- Earn money with each sale

Upload your text at www.GRIN.com
and publish for free

Bibliographic information published by the German National Library:

The German National Library lists this publication in the National Bibliography; detailed bibliographic data are available on the Internet at http://dnb.dnb.de .

This book is copyright material and must not be copied, reproduced, transferred, distributed, leased, licensed or publicly performed or used in any way except as specifically permitted in writing by the publishers, as allowed under the terms and conditions under which it was purchased or as strictly permitted by applicable copyright law. Any unauthorized distribution or use of this text may be a direct infringement of the author s and publisher s rights and those responsible may be liable in law accordingly.

Imprint:

Copyright © 2015 GRIN Verlag
Print and binding: Books on Demand GmbH, Norderstedt Germany
ISBN: 9783668756182

This book at GRIN:

https://www.grin.com/document/433459

Rainer Stickdorn

Ground water and contaminant flow modelling in Bremen area

GRIN Verlag

GRIN - Your knowledge has value

Since its foundation in 1998, GRIN has specialized in publishing academic texts by students, college teachers and other academics as e-book and printed book. The website www.grin.com is an ideal platform for presenting term papers, final papers, scientific essays, dissertations and specialist books.

Visit us on the internet:

http://www.grin.com/

http://www.facebook.com/grincom

http://www.twitter.com/grin_com

Bremen Modelling Project

Report of

Rainer Stickdorn

2nd semester MSc TropHEE
July 2015

Table of Contents

1 Executive Summary

In the northern german lowlands north west of Bremen and south west of river Weser a factory wants to extract 70 cubic meters of groundwater per hour. They want to pump clean water despite of a landfill as contaminant source south of the well (=upstream of the ground water flow). The well is very close to the western one of two parallel faults within the middle layer, with conductivities between 1 and 1E-9 m/s and is strongly influenced by them. The uppermost layer (= Saale) has the highest conductivity and is therefore most important for the flow regime. A MODFLOW model with 3 layers, given elevations and heads in the river and in the rim of the area was run many times to check if and how the well can can deliver the 70 cubic meters per hour of clean water.

Lowering pumping rate, displacement of the well away from the fault, … is only sufficient for the case of low conductivity (closed) faults. For open (high conductivity) faults these measures are not sufficient and a cleaning well north of the landfill is necessary. By trying lower and lower pumping rates of the cleaning well the minimum extraction rate for the cleaning well was found.

2 Project Description

2.a Introduction

The purpose of the groundwater model for an area south west of river Weser in Bremen is to check if and how a well in the midst of the model area can pump 70 cubic meters per hour for a new factory without getting contaminated water from a landfill site at the southern (lower) border of the area. The well is located at the more western one of two parallel faults in the middle layer running south west to north east (perpendicular to river Weser). The conductivity of these faults can vary between two extremes: 1 m/s for the open water filled fault and 1E-9 m/s for the closed fault filled with very fine sediments. Well location, depth and pumping rate can be changed a little bit. If unavoidable also a cleaning well can be placed to catch contaminated water. With ModelMuse, MODFLOW and other USGS-tools the flow of contaminants must be checked and visualised and a combination of parameters found that enables the factory to get clean groundwater in a suitable way. The task is described in more detail in [4a] and [4b].

2.b Site Description

Bremen is the capital of a northern westerly german state of the same name. It lies on both sides of the river Weser in the northern german flatlands 60 kilometres upstream of its estuary on the North Sea. Bremen has an elevation of about 11 meter a.s.l. Its geographic coordinates are 53° 5' N, 8° 48' E or in UTM projected coordinates the position is 32U 487091 5880729. The UTM coordinates are similar to the Gauss-Krüger coordinates that are given for the origin (= upper left corner, column 0, row 0) of the model area: x=3 478196.26, y= 5887455.54. The model area is 6.2 km (west-east) * 4.4 km (south-north) = 27.28 square km large (62 columns * 44 rows with 100m*100m wide cells) and mainly south west of the river Weser located. The area is located within the following left map [1]. The right one [6] indicates elevations:

Fig. 1 Map of Bremen	Fig. 2 Elevations in Bremen area
(Because of copy right reasons deleted)	(Because of copy right reasons deleted)

2.b.i Hydrologic Setting

The ground of the northern german lowlands close to North Sea is structured into at least 3 layers. The 2 uppermost are quarternary. The uppermost is Saale called after a tributary of Elbe river and one of the last glaciations between 130000 and 300000 years before present that left sediments of high conductivity Kx = 1 E-5 m/s.
The middle layer is Elster called after river Elster and an older glaciation between 300000 and 400000 years before present that left sediments of low conductivity kx = 1 E-5 m/s.
The lowest layer we deal with is tertiary i.e. older than quarternary (more than 2.6 mega years before present) and has a medium conductivity of kx = 1 E-4 m/s.

River Weser is flowing to the north-west like the rivers Ochtum (south of Weser) and Lesum (north of Weser) that join the river Weser north west of the area shown in the map. The inner city of Bremen lies on dunes and is the highest point in the area (yellow greenish colors in Fig. 2).

2.b.ii Site description – more data

Bremen has a moderate oceanic climate of classification Cfb (Köppen-Geiger). Winds are mainly westerly from the Atlantic Ocean. Mean annual precipitation is 671 mm (nearly exclusive rain). The annual mean temperature is 9.2 °C. For more detailed meteorological and climatic data see the following table ([2], [3]):

Climate data for Bremen

Month	Jan	Feb	Mar	Apr	May	Jun	Jul	Aug	Sep	Oct	Nov	Dec	Year
Record high°C (°F)	14.6 (58.3)	18.5 (65.3)	23.5 (74.3)	30.2 (86.4)	34.4 (93.9)	34.9 (94.8)	36.2 (97.2)	37.6 (99.7)	33.4 (92.1)	28.6 (83.5)	20.1 (68.2)	16.1 (61)	37.6 (99.7)
Average high °C (°F)	3.9 (39)	4.8 (40.6)	8.7 (47.7)	12.8 (55)	18.0 (64.4)	20.2 (68.4)	22.4 (72.3)	22.6 (72.7)	18.4 (65.1)	13.5 (56.3)	8.0 (46.4)	5.1 (41.2)	13.2 (55.8)
Daily mean °C (°F)	1.4 (34.5)	1.9 (35.4)	5.0 (41)	8.1 (46.6)	12.7 (54.9)	15.3 (59.5)	17.4 (63.3)	17.4 (63.3)	13.9 (57)	9.7 (49.5)	5.2 (41.4)	2.7 (36.9)	9.2 (48.6)
Average low °C (°F)	−1.1 (30)	−1.1 (30)	1.3 (34.3)	3.4 (38.1)	7.4 (45.3)	10.3 (50.5)	12.4 (54.3)	12.1 (53.8)	9.3 (48.7)	5.8 (42.4)	2.3 (36.1)	0.3 (32.5)	5.2 (41.4)
Record low °C (°F)	−21.8 (−7.2)	−23.6 (−10.5)	−18.7 (−1.7)	−7.6 (18.3)	−3.5 (25.7)	0.5 (32.9)	3.0 (37.4)	3.4 (38.1)	−1.2 (29.8)	−7.8 (18)	−14.1 (6.6)	−17.5 (0.5)	−23.6 (−10.5)
Average rain mm (inches)	55.1 (2.169)	35.6 (1.402)	51.2 (2.016)	40.8 (1.606)	54.2 (2.134)	73.4 (2.89)	65.0 (2.559)	61.2 (2.409)	60.1 (2.366)	55.4 (2.181)	57.7 (2.272)	61.6 (2.425)	671.3 (26.429)
Avg. rainy days ≥ 1 mm	11.3	8.6	11.0	9.0	9.5	11.1	10.8	10.1	10.6	10.5	11.5	12.0	126
Avg. relat. humidity %	87	84	80	75	71	73	75	75	81	84	87	88	80
Mean monthly sun hours	47	71	107	170	214	193	205	193	143	108	54	40	1,545

Fig. 3 Climate data for Bremen

River Weser has a mean discharge of 337 cubic meters per second.

3 Data Assimilation

Geological and hydrogeological data from around 70000 bore holes in Bremen and Bremerhaven is available from "Geologischer Dienst für Bremen – GDFB" (https://www.gdfb.de/).

For the MODFLOW-model a text file (3_Layers_DATA.txt) as digital elevation model – DEM with geometric z-coordinates in 62*44 cells in each of 3 layers as point data and a shapefile (GW_Boundary_GK3.shp) with hydraulic heads for all 210 cells at the model rim (2*44 + 2*62 -2*1 = 2* (44+62-1) = 2*105 = 210) were given instead of any recharge data or heads or groundwater levels in single observation wells. Analysis of bore hole data was not necessary, because geometry of the layers (see above) and conductivities (highest k=1E-3 in top layer = Saale, lowest K=1E-5 in middle layer = Elster, intermediary k=1E-4 in lowest layer = tertiary, special values of 1 and 1E-9 for fault zone where Elster layer is less than 4m thick) were explicitly given. The x- and y-coordinates of the river were given as shapefile "Weser_AOI_DHDN3_z3.shp". The river bed conductance K=1E-5 m/s, thickness M=3 m, cell length L=100m and width W=100m was explicitly given, so that a constant C=k*L*W/M = 0.1m^2/3sec = 0.0333 m^2/sec as Direct-Value for the RIV-module was determined.

The point elevation file "3_Layers_DATA.txt" should have been prepared in a way, that no discrepancies like elevation (lowest layer) > elevation (middle layer) i.e. negative thickness of middle layer are found. But 3 points were overlooked and I adjusted them to elevation (lowest layer) = elevation (middle layer) – 1m before loading it in ModelMuse with File-Import-Points as a single object.

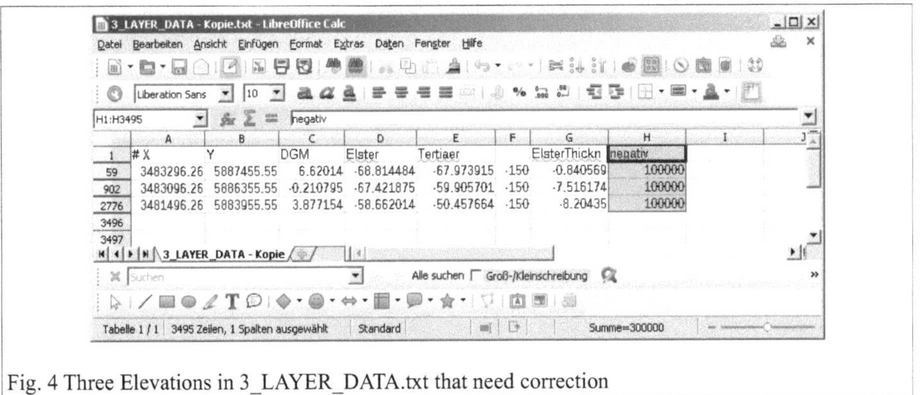

Fig. 4 Three Elevations in 3_LAYER_DATA.txt that need correction

4 Description of the groundwater model

Groundwater modeling is described in [15] and in many publications of USGS. Some of them can be found in the References section of this paper.

4a Spatial discretization

In ModelMuse a model with 62 columns (column 1 = west border, column 62 = east border), 44 rows (row 1 = north border, row 44 = south border) both of 100m width, 3 layers and origin at x=3478196.26 and y=5887455.54 (easting and northing in meter in Gauss-Krüger zone 3 or DHDN3 around longitude of 3*3°=9°E) was created. This area represents 6.2 km * 4.4 km = 27.28 square km north west of the city of Bremen. The geometric z-values of Model_Top, Upper_Aquifer_Bottom = Saale_Bottom, Middle_Aquifer_Bottom = Elster_Bottom and Lower_Aquifer_Bottom = Tertiary_Bottom were imported as point text file "3_Layers_DA-TA.txt". The layers were renamed according to the description in chapter 4.b. The first layer (uppermost layer) was renamed to Saale. The second (middle) layer is Elster and the third (lowest) layer Tertiary. The horizontal spreading of river Weser was imported as shapefile "Weser_AOI_DHDN3_z3.shp". The hydraulic / piezometric heads in all cells at the very rim of the model area were imported as (the z-values in) shapefile "GW_Boundary_GK3.shp" for the GHB-Module.

All cells are active. The layers have no further discretization (Vertical discretization = 1 in MODFLOW Layer Groups) and are of type Confined. The Confined-type was selected as described and justified in [4b]. The related storitivity S uses the default value of 1E-5.

4b Temporal discretization

Under Model-MODFLOW-Time "Steady State" was selected (1 stress period from -1 to 0), i.e. no time discretization.

4c Heterogeneity and layering

The layers were marked as rectangular objects each with 2 z-values equal to layer borders and got horizontal conductivities Kx of 1E-3 (Saale), 1E-5 (Elster if thicker than 4m or 1 or 1E-9 in other case which represents the fault), 1E-4 (Tertiary):

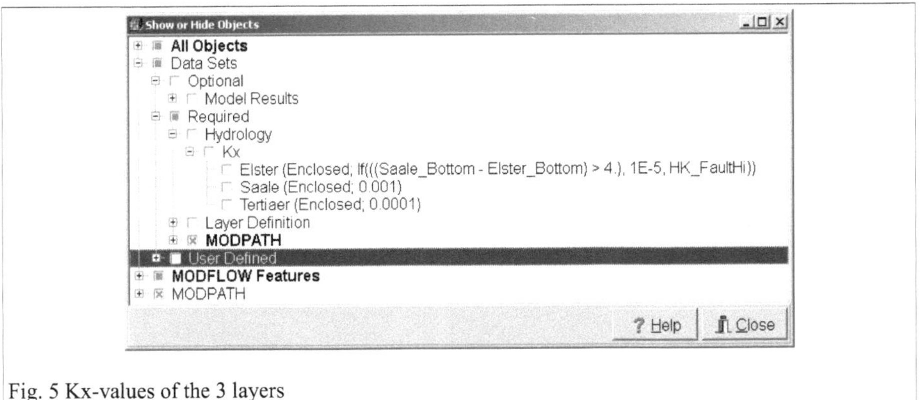

Fig. 5 Kx-values of the 3 layers

For the conductivity Kx within the faults two global Variables HK_FaultHi=1 and HK_FaultLo = 1E-9 are used. The formula "If(...)" does not need the Abs()-Function for the difference of Layer-z-values, because for all points was checked that Tertiary has a lower elevation than Elster and discrepancies (only 3 points) were corrected before importing the text-file. Vertical conductivity is Ky=0.1 * Kx as default.

Visualizing Kx for the middle layer (= Elster) makes the 2 faults visible (2 parallel red lines from lower left to upper right). They are perpendicular to the river Weser that flows from lower right to upper left corner. A well at x=3480869 (east) and y=5885587 (meters north of equator) i.e. in the cell at column 27 and row 19 is found at the western rim of the western fault. A landfill site (= contamination source) is found at x=348437 and y=5883148 i.e. in the cell at column 23 and row 44 at the southern rim between the 2 faults:

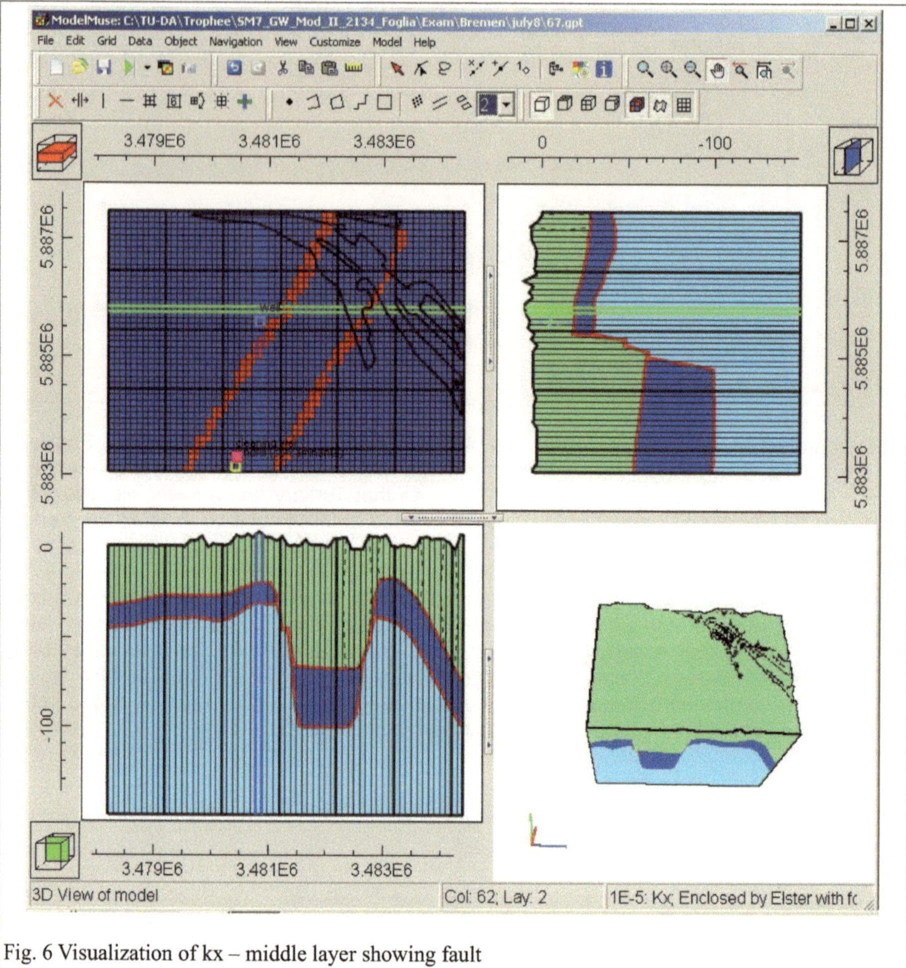

Fig. 6 Visualization of kx – middle layer showing fault

4d Solver

The default solver PCG (Preconditioned Conjugate Gradient package) with default parameters was used.

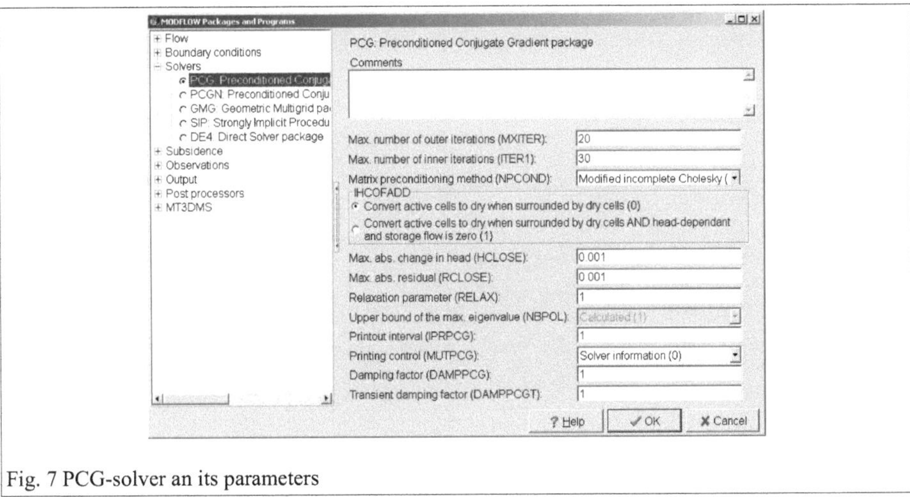

Fig. 7 PCG-solver an its parameters

In the first but complete description of MODFLOW ([7]) the iterative numeric solution of Finite-Difference equations and the very first two solvers (Strongly Implicit Procedure Package – SIP and Slice-Successive Overrelaxation Package – SSOR) are described. The new default solver PCG is described in chapter 7 of [8]. PCG uses outer and inner iterations that are limited by the two parameters MXITER and ITER1 (here with default values 20 and 30). Inner iterations stop earlier if convergence is reached: change in heads between itera-tions < HCLOSE = 0.001, otherwise a new outer iteration is started. The matrix precon-ditioning method (NPCOND) is here "Modified Imcomplete Cholesky" which is faster on classical PCs. The alternative "Polynomial" is only recommended for parallel computers (many microprocessors or GPU-parallel-processors). Cholesky is a matrix decomposition into a product of simplier matrixes that makes numerical solution of the equation system easier.

4e Boundary Conditions

Concepts for boundary conditions are described in detail in [13] and [14]. For boundary con-ditions no package under "Specified head", but the WEL-package under "Specified flux" and the packages GHB (General Head Boundary) and the RIV-package under "Head-dependent flux" were configured. Stream-, Recharge- and Evapotranspiration-Packages were not to be used.

The WEL-package configures the well in position (Gauss-Krüger coordinates) x=3480869 (easting) and y=5885587 (northing) that is located in the cell in column 27 and row 19. To avoid pumping contaminated water also well positions up to several hundred meters (up to 3 diagonal cells) to the north west away from the western fault were tested. The well has a pumping rate of 70 cubic meters per hour or 70/3600 = 0.01944 cubic meters per second. Instead of this 100% pumping rate also 50% and 25% of this value were tried in case of contaminants in the well in the aim to avoid them. Also different well depths (only layer 1 or layer 1 to 3 or only layer 3) were tested. The well package is described in detail in chapter 8 of [7] and shortly in chapter 6 and 8 of [8]:

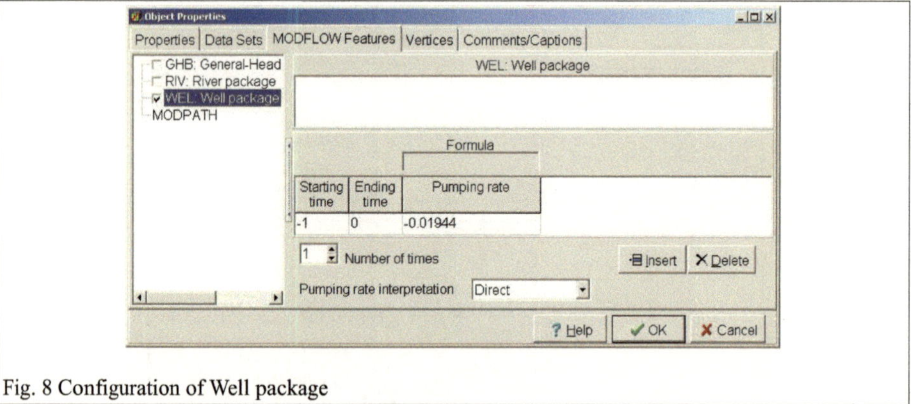

Fig. 8 Configuration of Well package

As the upper layer (Saale) has the highest conductivity (100 times higher than the middle layer) drilling into deeper layers should normally not be usefull, because almost all water can be expected to came out of the top layer even if the well reaches much deeper and drilling deeper is more expensive.

The GHB-Modul is described in detail in chapter 11 of [7] and shortly in chapter 6 and 8 of [8]. It was configured with the imported shapefile "GW_Boundary_GK3.shp" that sets the piezometric heads in all 210 (2*(62 columns + 44 rows -1 for saving at corner) = 2*105 = 210) rim cells to its z-value. Conductivity kx is inherited from layer properties.

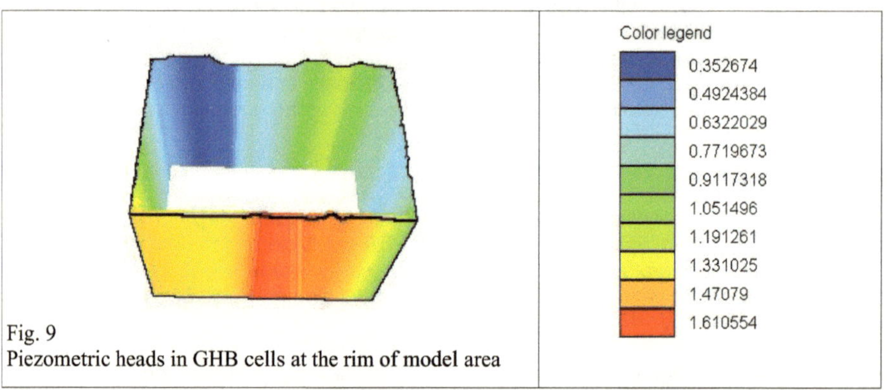

Fig. 9
Piezometric heads in GHB cells at the rim of model area

At the southern border the piezometric heads are highest (red) and at the northern border they are lowest (blue). So we can expect a flow from south to north that will hit the river Weser. We have bad luck with the landfill site (= contamination source) that is also at the southern rim close to the zone of high heads. So the groundwater flow from south to north will take contaminants with it and potentially transport these contaminants to the well and to the river.

The RIV-package is described in detail in chapter 6 of [7] and in chapter 6 and 8 of [8]. Package RIV is configured concerning the x- and y-extension of the Weser by the shapefile

"Weser_AOI_DHDN3_z3.shp". Also the River stage is taken from the z1-values of the shape-file. River bottom is just 9 m less. River bed thickness M=3m, -conductivity K=1E-5 m/s, width of river W=100m and cell length L=100m result in a calculated river Conductance C = K*W*L/M = 1E-5 m/s * 100m * 100m / 3m = 0.1 m^2 /3s = 0.033 m^2 /sec.

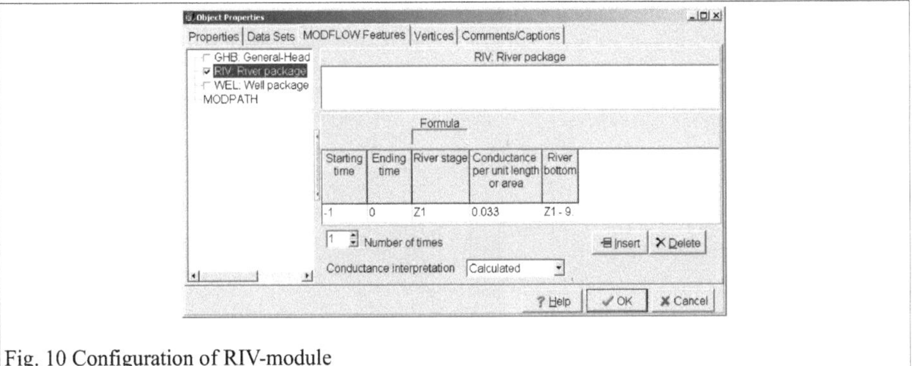

Fig. 10 Configuration of RIV-module

The landfill site at x=3480437, y=5883148 i.e. in cell at column 23 and row 44 is configured as source of particles for MODPATH. MODPATH is a postprocessor to MODFLOW. MODPATH is described in [10]. It uses the default porosity of 0.25. The contamination source is restricted to layer 1 (Saale) that is there around 50 m thick and underlain by a layer of much smaller conductivity.

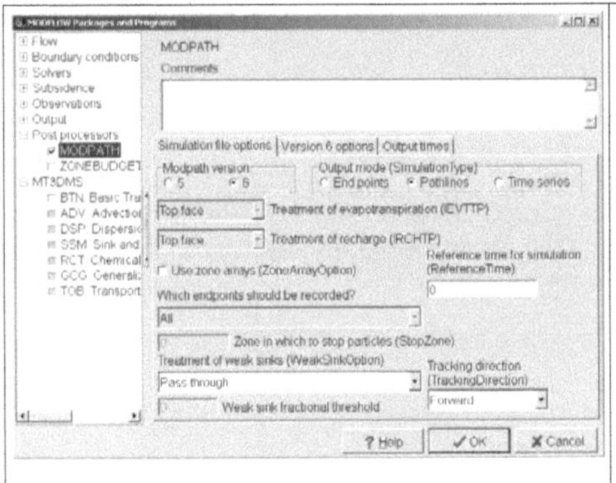

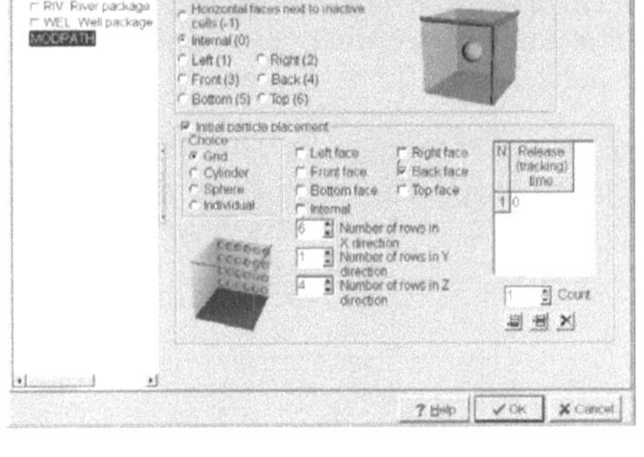

Fig. 11: MODPATH (postprocessor) generates Flow lines / Path lines of particles

Fig. 12: landfil site (contamination source) configured as source of particles for MODPATH

For the cases where contaminants can not be kept from the factories well by other means a cleaning well is put into the cell at column 23 and row 42 i.e. just north of the landfill and operated with different pumping rates to extract contaminated water.

5 Discussion of Flow Model Results

Regardless of the parameters we can change (well position, well depth, pumping rate, pumping rate of cleaning well, ...) to get rid of contaminants in the well we always have to distinguish two cases of different kx-values for the faults: I used 2 global variables HK_FaultHi = 1 for open faults and HK_FaultLo = 1E-9 for closed faults. For the low conductivity faults (i.e. closed faults) and without any pumping the heads and flowlines are the following (run 37 in the table in chapter 8). For producing flow/pathlines with ModelMuse southwest of the river several point objects were placed at the western and southern border of the model area and configured for MODPATH by putting lots of particles into them:

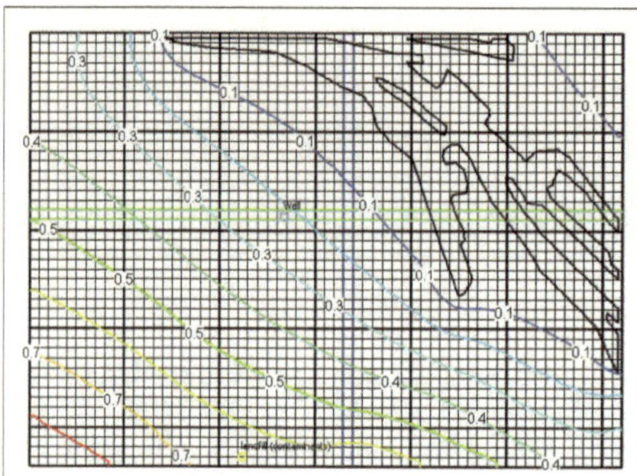

Fig. 13: Heads for low conductivity (=closed) faults & no pumping

Fig. 14: Flow for low conductivity (=closed) faults & no pumping

Volumetric budget from *.lst-file:
Into aquifer: from head dependent boundaries 0.1447 m^3/s
out of aquifer: into river (= river leakage) 0.1437 m^3/s
In-Out 9.9 E-4 = 0.7 % discrepancy

USGS-Tool GW_Chart (found in [12]) is a program for Calibration Plots, Water Budged Plots, Hydrographs, Lake plots, Piper diagrams and Cell Water Budgets. It reads for example MODFLOW's lst-files. Here it is used to plot the water balance:

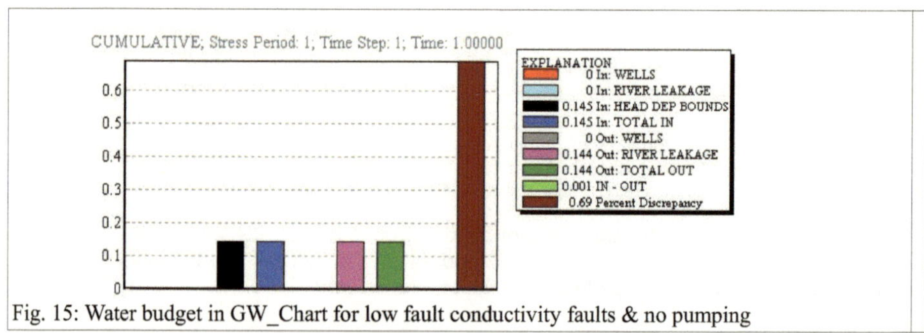

Fig. 15: Water budget in GW_Chart for low fault conductivity faults & no pumping

For high conductivity (= open) faults and without any pumping the heads and flowlines are the following (run 36 in the table in chapter 8).

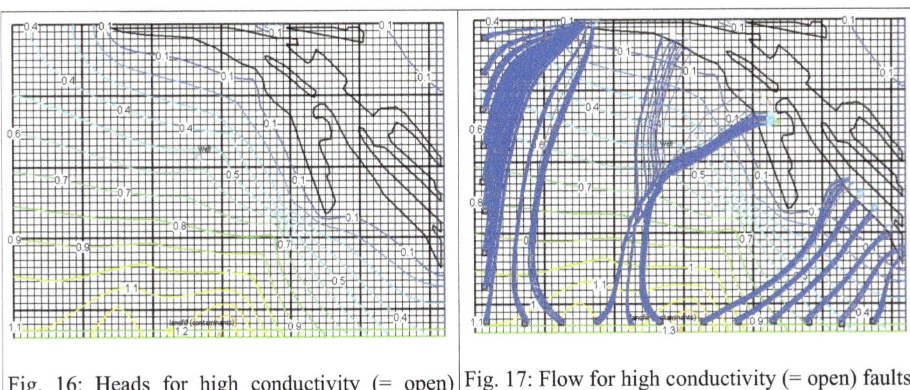

Fig. 16: Heads for high conductivity (= open) faults & no pumping

Fig. 17: Flow for high conductivity (= open) faults & no pumping

With the high conductivity (Kx=1) in the faults flow becomes more heterogenious and is devited into 3 areas: west of the faults, fault area and east of the faults. Also much more water is flowing in northern and north eastern direction into the river.

Volumetric budget from *.lst-file:
Into aquifer: from head dependent boundaries 0.5835 m^3/s
out of aquifer: into river (= river leakage) 0.5777 m^3/s
In-Out 5.75 E-3 = 1 % discrepancy

USGS-Tool GW_Chart plots the following water balance:

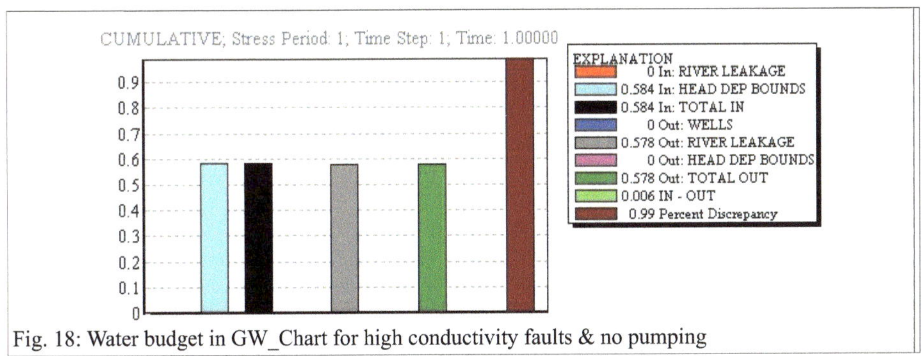

Fig. 18: Water budget in GW_Chart for high conductivity faults & no pumping

Running the well in cell in column 27 and row 19 with 100% (= -0.01944 m^3/sec) with low (run 33) und high (run 32) conductivity faults:

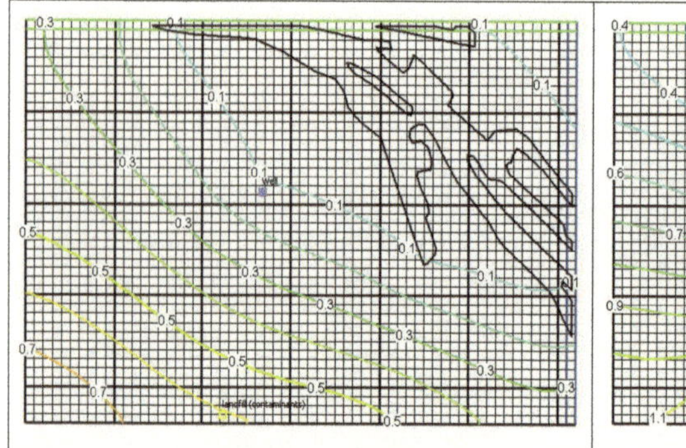 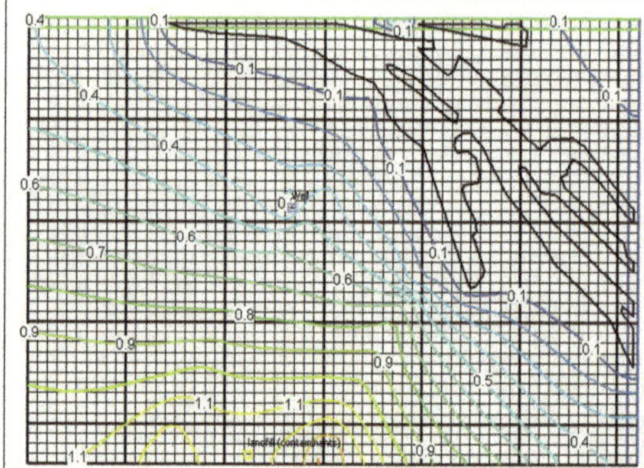

Fig. 19: Heads for low conductivity (closed) faults & 100% pumping (# 33)	Fig. 20: Heads for high conductivity (open) faults & 100% pumping (# 32)

Flow direction and strength is shown with vectors for the head gradient in Model Viewer. Model Viewer is for 3-D-visualization of model results and described in [11]. The vectors show direction and strength of flow:

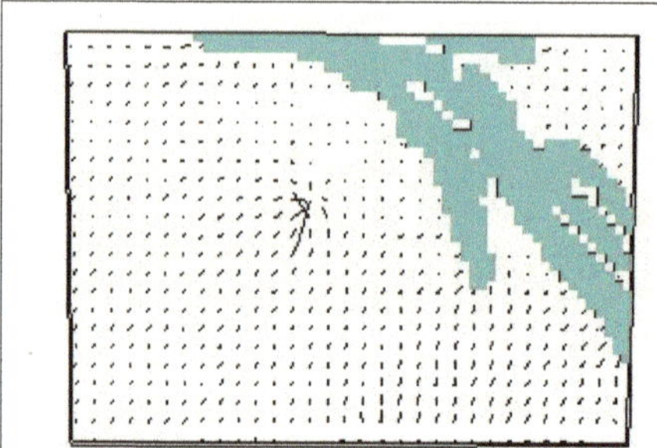 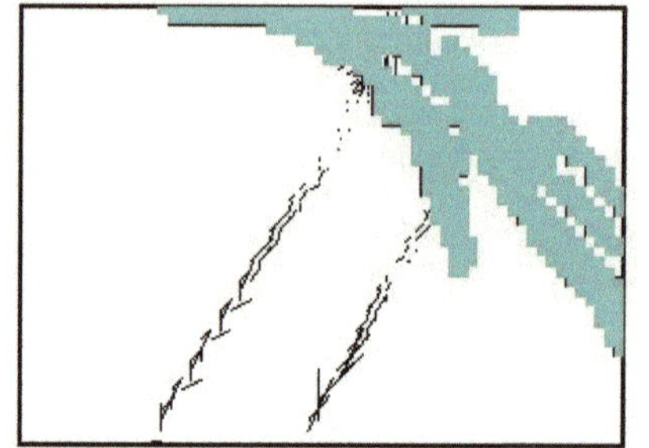

Fig. 21: Vectors (gradient of heads) for low conductivity faults & 100% pumping	Fig. 22: Vectors (gradient of heads) for high conductivity faults & 100% pumping

Both pictures use a Scale Factor of 700000 within Model Viewer. With low conductivity faults the well as "focal point" is easily seen. With high conductivity faults, flow is concentrated into the faults. As the well is very close to the left fault it disappears as focal point. Big amounts of water flow through the fault so the little amount the well extracts has no visible effect. Also all the small vectors in other parts of the area are so small compared to that at the fault, that a huge Scale Factor would be needed to make them visible. GW_Chart shows the water budget for both cases with much more water moved in the case of high conductivity (open) faults:

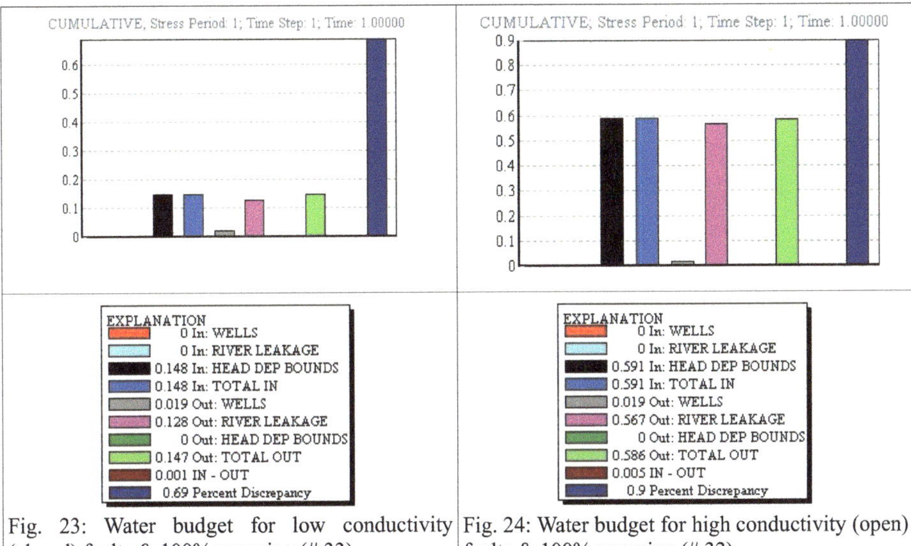

Fig. 23: Water budget for low conductivity (closed) faults & 100% pumping (# 33)

Fig. 24: Water budget for high conductivity (open) faults & 100% pumping (# 32)

The factory wants to pump clean water from the well. We must check if contaminants from the landfill at the southern border (cell in column 23, row 44, layer 1) are transported by advection into the well. The landfill point object is therefore filled with particles for MODPATH to track:

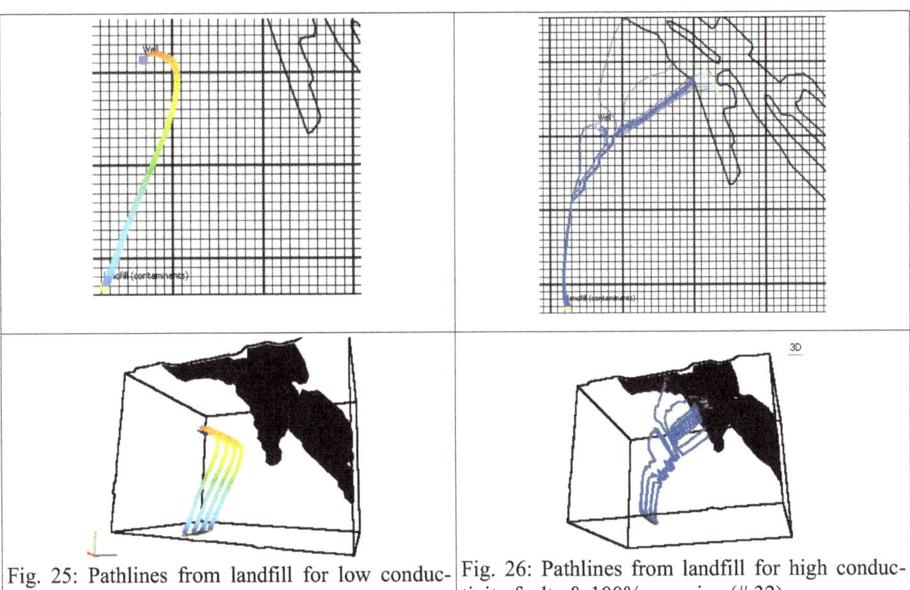

Fig. 25: Pathlines from landfill for low conductivity faults & 100% pumping (# 33)

Fig. 26: Pathlines from landfill for high conductivity faults & 100% pumping (# 32)

In both cases the well is contaminated with material from the landfill. So different scenarios changing well location, depth, pumping rate, … must be tried to avoid contamination of the well.

6 Description of scenarios

The different combinations of well position, well depth, pumping rate, cleaning well pumping rate and fault conductivity are documented in a more compact format as a table.

- The 1st column contains the number of the program run (= name of the files like 32.nam, 32.path, 32.lst, …).
- The 2nd column indicates the fault conductivity: Hi=1, Lo=1E-9.
- The 3rd column contains the location of the well: the original position (col 27, row 19) is changed up to 3 cells diagonal to north west (3*100m*1.41= 424m) i.e. away from the western fault.
- The 4th column shows the pumping rate: 100% = -0.0194 cubic meters/sec.
- The next 2 columns show if the contaminant plume hits the well and/or the river.
- The 3rd column from right shows the water budget with InGHB = amount that flows from the GHB into the aquifer, OutWell = amount that flows out of the aquifer by wells (production well + cleaning well), OutRiv = amount that flows out of the aquifer into the river. The balance is: InGHB = OutWell + OutRiv.
- The last but one column shows if a cleaning well is used and in which cell it is positioned.
- The last column contains the pumping rate of the cleaning well in cubic meters per second.

Green rows show combinations without a plume in the production well with a pumping rate of at least 25% of the 70 cubic meters per hour (= 0.0194 cubic meters / second).

#	Fault-Kx	Pump Loc Col/Row/Lay	Pump rate%	Plume in well	Plume in river	InGHB/OutWell/Out Riv	Clean Col/Row/La	Clean rate
32	Hi	27/19/1	100	yes	yes	0.5912/0.0194/0.5665	no	no
33	Lo	27/19/1	100	yes	No	0.1484/0.0194/0.1279	no	no
34	Lo	27/19/1-3	100	yes			no	no
35	Hi	27/19/1-3	100	yes			no	no
36	Hi	27/19/1	0	yes		0.5835 / - / 0.5777	no	no
37	Lo	27/19/1	0	No	Yes	0.1447 / - / 0.1437	no	no
38	Lo	26/18/1	100	yes			No	no
39	Hi	26/18/1	100	yes			No	No
40	Hi	26/18/1	50	yes			No	No
41	Lo	26/18/1	50	No	Yes	0.1467/ 0.01/0.1356	No	No
42	Hi	26/18/1	25	yes			No	No
43	Hi	26/18/1	0	yes			No	No
44	Hi	24/16/1	0	yes			No	No

#	Fault-Kx	Pump Loc Col/Row/Lay	Pump rate%	Plume in well	Plume in river	InGHB/OutWell/Out Riv	Clean Col/Row/La	Clean rate
45	Hi	24/16/3	0	No?1)	Yes	0.584/ 0 / 0.578	No	No
46	Hi	24/16/3	25	Yes			No	No
47	Hi	24/16/3	50	Yes			No	No
48	Hi	24/16/1	100	No	No	0.5984/0.0294/0.5634	23/42/1	-0.01
49	Hi	24/16/1	100	No	No	0.5940/0.0244/0.5650	23/42/1	-0.005
50	Hi	24/16/1	100	Yes	Yes	0.5922/0.0219/0.5650	23/42/1	-0.0025
51	Hi	24/16/1	100	Yes	Yes	0.5927/0.0224/0.5648	23/42/1	-0.003
52	Hi	24/16/1	100	No	No	0.5935/0.0234/0.5647	23/42/1	-0.004
53	Lo	24/16/1	50	No	Yes	0.1467/ 0.01/0.1357	23/42/1	0
54	Lo	26/18/1	50	No	Yes	0.1467/ 0.01/0.1356	23/42/1	0
55	Lo	26/18/1	75	No	Yes	0.1475/0.0146/0.1319	23/42/1	0
56	Lo	24/16/1	25	No	Yes	0.1457/0.005/0.1397	23/42/1	0
57	Lo	27/19/1	50	No	Yes	0.1466/ 0.01/0.1356	23/42/1	0
58	Lo	27/19/1	25	No	Yes	0.1457/0.005/0.1397	23/42/1	0
59	Lo	27/19/1	100	Yes	No	0.1496/0.0219/0.1267	23/42/1	-0.0025
60	Lo	27/19/1	100	Yes	No	0.1498/0.0224/0.1265	23/42/1	-0.003
61	Lo	27/19/1	100	No	No	0.1503/0.0234/0.1260	23/42/1	-0.004
62	Lo	27/19/1	100	No	No	0.1501/0.0229/0.1262	23/42/1	-0.0035
63	Lo	27/19/1	75	No	Yes	0.1475/0.0146/0.1319	23/42/1	0
64	Hi	27/19/1	100	No	No	0.5952/0.0244/0.5657	23/42/1	-0.005
65	Hi	27/19/1	100	No	No	0.5943/0.0234/0.5661	23/42/1	-0.004
66	Hi	27/19/1	100	Yes	Yes	0.5936/0.0224/0.5659	23/42/1	-0.003
67	Hi	27/19/1	100	Yes	Yes	0.5939/0.0229/0.5662	23/42/1	-0.0035
68	Hi	24/16/1	100	Yes	Yes	0.5932/0.0229/0.5646	23/42/1	-0.0035
69	Lo	24/16/1	100	No	No	0.1515/0.0234/0.1261	23/42/1	-0.004
70	Lo	24/16/1	100	2)		0.1515/0.0234/0.1261	23/42/1	-0.004
71	Hi	24/16/1	100	2)		0.5931/0.0234/0.5655	23/42/1	-0.004

1) Contaminants seem to flow directly over the well (not into the well)
2) Additional particle source placed between both wells for checking release from old contaminated soil (contaminated before cleaning well was installed)

6.1 Scenarios without cleaning well

Open faults (with high conductivity):
Even if we drop the pump rate to 0% we get contaminants into the production well (see # 36, 43, 44), also after moving the well some hundred meters to north west away from the western

fault. Only if we restrict the screen of the well to layer 3 (see # 45) we don't get contaminants into the well at pumping rate 0%. Even with a little pumping of 25% (see # 46) contaminants are sucked into the well screen in layer 3. Moving the well, restricting its screen to lower level and reducing the pump rate (but 25% is not acceptable) we can't provide the factory with clean water.

Closed faults (with low conductivity):
Program run # 41 provides clean water (no plume in well) at pumping rate 50%. Run # 55 (well relocated to north west) and run # 63 (original well location) was even successful for 75% pumping rate. 100% pumping rate causes in all investigated cases a plume in the well. Only if the factory agrees to get only 75% of the 70 cubic meters per hour we can deliver clean water in the case of closed faults without a cleaning well.

6.2 Scenarios with a cleaning well

If we put a cleaning well into the cell at column 23, row 42, layer 1 that is north of the contaminants clean water is more easily provided for the factory. To save energy the cleaning pump should run with the lowest pumping rate possible. As in 8.1 the open faults are the more problematic ones.

Open faults (with high conductivity):
Program run # 52 (well relocated to north west) and # 65 (well in original position) show, that we can get clean water at the production well at a pumping rate of 100% as long as the cleaning well pumps at least with a rate of 0.004 cubic meters per second (4 liters/sec).

Closed faults (with low conductivity):
Program run # 62 shows that clean water at the production well running at 100% is provided if the cleaning well pumps at least 0.0035 cubic meters per second (3.5 liters/sec).

7 Conclusion

Without a cleaning well it´s difficult to deliver clean water to the factory. Only in case of closed faults a decent part (75%) of the water that is asked for can be delivered. In case of open faults this is not possible.

So I propose to install a cleaning well in the northern vicinity of the contaminated side and to pump with at least 4 liters per second which is sufficient for both conductivities of the fault to catch all contaminants and to prevent them from traveling to the production well (or the river). Both wells are limited to the first = upper layer (Saale) because its conductivity is 100 times larger than that of the underlying layer. Water and therefore also contaminant flow is concentrated to the upper layer.

I also propose to relocate the well some hundred meters to the north west – away from the western fault to safeguard the filter function of the aquifer. An open fault can be a shortcut between (new) contamination spots in the south or organic pathogen sources south of the production well. If the water travels more than 50 days (water protection zone II criteria) through the aquifer contained organic pathogens are normally death and therefore hazard-free. A dislocation of the production well of more than same hundred meters seems unrealistic since it should remain on the grounds of the factory.

The recommendation is realized in program run # 52 (high conductivity) and # 69 (low conductivity). Following pictures show the head distribution as contour plots in ModelMuse:

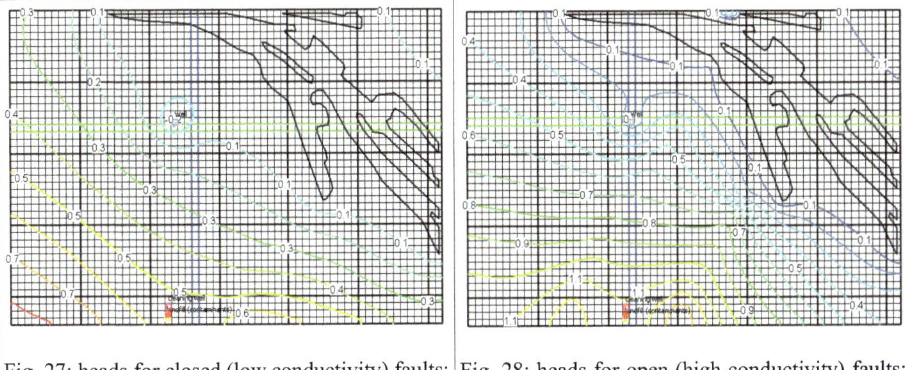

Fig. 27: heads for closed (low conductivity) faults; 100% pumping rate, cleaning well w 4 L/sec (#69)

Fig. 28: heads for open (high conductivity) faults; 100% pumping rate, cleaning well w 4 L/sec (#52)

Model Viewer shows the heads color coded and the (short) pathlines between landfill and cleaning well after loading the files 69.bhd, 52 bhd, 69.path and 52.path:

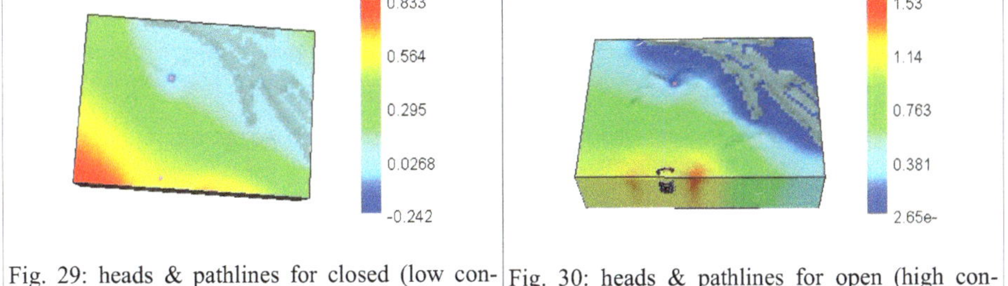

Fig. 29: heads & pathlines for closed (low conductivity) faults (run # 69)

Fig. 30: heads & pathlines for open (high conductivity) faults (run # 52)

The water budget for both cases produced by GW_Chart by reading the files 69.lst and 52.lst:

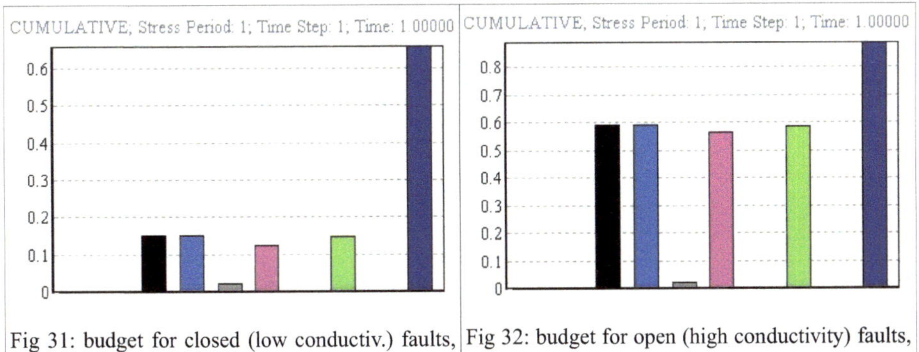

Fig 31: budget for closed (low conductiv.) faults,

Fig 32: budget for open (high conductivity) faults,

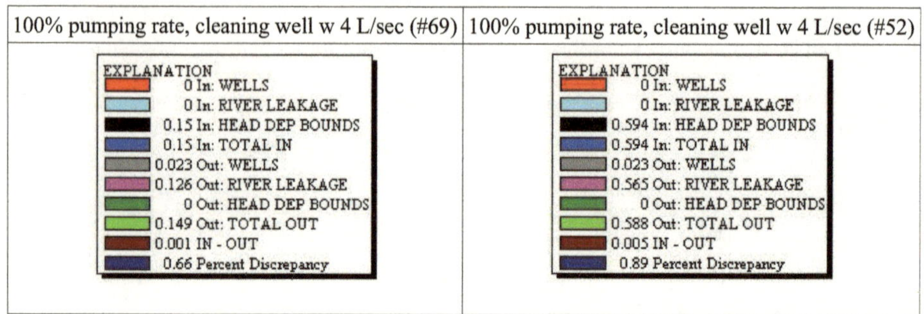

100% pumping rate, cleaning well w 4 L/sec (#69)	100% pumping rate, cleaning well w 4 L/sec (#52)

8 Recommendation for further work

8.1 Faults

The faults conductance was assumed to be either 1 or 1E-9 m/s. These values are extremes. The real faults will vary between these values depending on location. As the program runs without a cleaning well have shown there are big differences in head distribution, head gradients (vector diagrams), water budgets and the question if and when a plume will reach the production well, the nature of the faults should be known better.

To get more realistic data, some observation wells should be build between the southern border of the modeling area and the river to test for contaminants, heads, conductivities, ...

Such observations could also be used to check the model and to adjust model parameters until the model fits better the observations.

8.2 Contaminants

Even if the cleaning well directly north of the landfill catches all contaminants that are newly released from the landfill there is probably already a lot of contaminated soil on the path to the north and a large part of these contaminants will continue their travel to production well and river. We can check that with a point object with contained MODPATH particles e.g. half way between production and cleaning well:

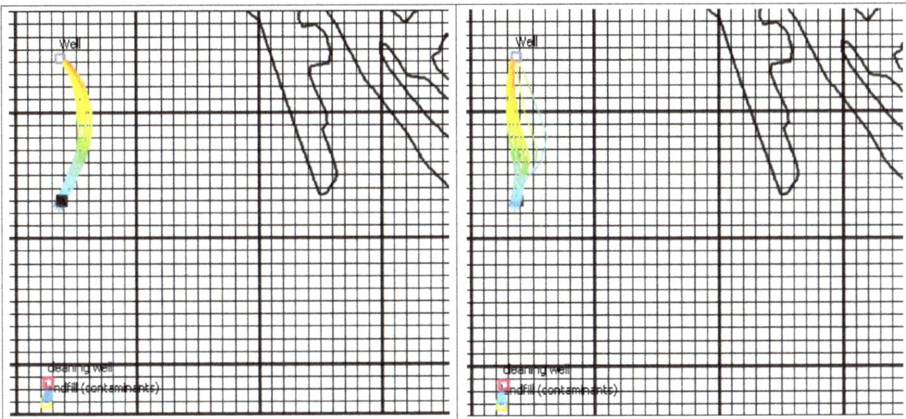

Fig. 33: Continued polution despite functioning cleaning well for low conductivity faults (# 70)

Fig. 34: Continued polution despite functioning cleaning well for high conductivity faults (# 71)

It has to be investigated how long the water from the production well will contain contaminants that are not coming directly from the landfill but from soil between production well and landfill that was contaminated earlier (before cleaning well started working) and how high the contamination will be. For how long must water from production well be discarded?

8.3 Seasonal data

The data given was not related to seasons. GHB-data as annual means are a hefty simplification. At least it should be checked if this data changes between seasons. As Fig. 3 shows there are big differences in precipitation, temperature, sun shine hours, ... between different months. Perhaps a more elaborated model with parameters depending on season is necessary.

9 References

[1] Coordinate Calculator converting between Gauss-Krüger, UTM, … geographic coordinates
http://www.deine-berge.de/Rechner/Koordinaten/Dezimal/53.118844,8.6732

[2] World Meteorological Organisation – WMO – and german weather service – DWD - data for Bremen: http://worldweather.wmo.int/en/city.html?cityId=1342

[3] Wetterkontor (weather office) with meteorological data for Bremen
http://www.wetterkontor.de/de/klima/klima2.asp?land=de&stat=10224

[4a] Foglia, L., 2015: "Introduction to Groundwater Modeling – Model Exercise (Bremen)", Institut of Applied Geology, TU-DA, 12 pp.

[4b] Schuhmann, P., 2015: "Construction of a groundwater model with MODFLOW-2005 by using data processed in SKUA", Institut für Angewandte Geowissenschaften, TU-DA, (file "Construction of … in SKUA_forexam.docx"), 17 pp.

[5] European Environment Agency – Digital Elevation Model over Europe: EU-DEM in ERTS89 http://www.eea.europa.eu/data-and-maps/data/eu-dem#tab-original-data

[6] Bremen, Germany, Elevation Map
http://www.floodmap.net/Elevation/ElevationMap/?gi=2944388

[7] McDonald, M.G., Harbaugh, A.W., 1988: "A Modular three-dimensional Finite-Difference Ground-water Flow Model" - Techniques of Water-Resources Investigations -TWRI - Book 6 Modeling Techniques, USGS, United States Government Printing Office, 600 pp.

[8] Harbough, A.W., 2005: "MODFLOW-2005, The U.S. Geological Survey Modular Ground-Water Model-the Ground-Water Flow Process" - Book 6. Modeling techniques (TM6A16), USGS, 253 pp.

[9] Winston, R.B., 2009: "ModelMuse – A Graphical User Interface for MODFLOW-2005 and PHAST.", USGS (TM6A29), 59 pp.

[10] Pollock, D.W., 2012: "User Guide for MODPATH Version 6 – A Particle-Tracking Model for MODFLOW", USGS (TM_6A_41), 70 pp.

[11] Hsieh, P.A., Winston, R.B., 2002: "Users's Guide to Model Viewer, A program for three-dimensional visualization of ground-water model results", USGS (Open-File Report 02-106), 23 pp.

[12] Water Resources Groundwater Software from USGS
http://http://water.usgs.gov/software/lists/groundwater/

[13] Franke, O.L., Reilly, T.E., Bennett, G.D., 1987: "Definition of Boundary and initial conditions in the analysis of saturated ground-water flow systems – An introduction", Techniques of Water-Resources Investigations – TWRI - of USGS, Book 3 chapter B5, 22 pp.

[14] Reilly, T.E., 2001: "System and Boundary Conceptualization in Ground-Water Flow Simulation", Techniques of Water-Resources Investigations -TWRI - of USGS, Book 3, Chapter B8, 38 pp.

[15] Anderson, M.P., Woessner, W.W., 1992: "Applied Grundwater Modeling – Simulation of Flow and Advective Transport", Academic Press, Elsevier Science Publishing, 381 pp.

10 Table Of Figures

Figure	Title	Source	page
1	Map of Bremen	[1]:http://www.deine-berge.de/Rechner/Koordinaten/Dezimal/53.118844,8.6732	3
2	Elevation in Bremen area	[6]:http://www.floodmap.net/Elevation/ElevationMap/?gi=2944388	3
3	Climate Data for Bremen	[2]:http://worldweather.wmo.int/en/city.html?cityId=1342, [3]:http://www.wetterkontor.de/de/klima/klima2.asp?land=de&stat=10224	4
4	Three Elevations in 3_LAYER_DATA.txt that need correction	Authors Libre Office Calc	5
5	Kx-values of the 3 layers	Parameters in USGS-MODFLOW	6
6	Visualization of kx – middle layer showing fault	Graphs within USGS-Model Muse	7
7	PCG-solver an its parameters	Solver parameters within USGS-MODFLOW	8
8	Configuration of Well package	Well parameters within USGS-MODFLOW	9
9	Piezometric heads in GHB cells at the rim of model area	Authors modelling results for the area border from USGS-MODFLOW	9
10	Configuration of RIV-module	River parameters within USGS-MODFLOW	10
11	MODPATH (postproc.) generates Flow lines / Path lines of particles	Authors parameter selection within USGS-MODFLOW	10
12	landfil site (contamination source) configured as source of particles for MODPATH	Authors parameter selection for USGS-MODPATH	10
13	Heads for low conductivity (=closed) faults & no pumping	Authors modelling results in USGS-MODPATH for water levels	11
14	Flow for low conductivity (=closed) faults & no pumping	Authors modelling results in USGS-MODPATH for water flows	11
15	Water budget in GW_Chart for low fault conductivity faults & no	Authors modelling results in USGS-GW CHART for water budget	11

Figure	Title	Source	page
	pumping		
16	Heads for high conductivity (= open) faults & no pumping	Authors modelling results in USGS-MODPATH for water levels	12
17	Flow for high conductivity (= open) faults & no pumping	Authors modelling results in USGS-MODPATH for water flows	12
18	Water budget in GW_Chart for high cond. faults & no pumping	Authors modelling results in USGS-GW CHART for water budget	12
19	Heads for low conductivity (clos.) faults & 100% pumping (# 33)	Authors modelling results in USGS-MODPATH for water levels	13
20	Heads for high conductivity (open) faults & 100% pumping (# 32)	Authors modelling results in USGS-MODPATH for water levels	13
21	Vectors (gradient of heads) for low cond. faults & 100% pumping	Authors modelling results in USGS-Model Viewer	13
22	Vectors (gradient of heads) for high cond. faults &100% pumping	Authors modelling results in USGS-Model Viewer	13
23	Water budget for low conductivity (closed) faults & 100% pumping (# 33)	Authors modelling results in USGS-GW CHART for water budget	14
24	Water budget for high cond.(open) faults & 100% pumping (# 32)	Authors modelling results in USGS-GW CHART for water budget	14
25	Pathlines from landfill for low conductivity faults & 100% pumping (# 33)	Authors modelling results in USGS-MODPATH	14
26	Pathlines from landfill for high conductivity faults & 100% pumping (# 32)	Authors modelling results in USGS-MODPATH	14
27	heads for closed (low cond.) faults;100% pumping rate, cleaning well w 4 L/sec (#69)	Authors modelling results in USGS-Model Muse	18
28	heads for open (high cond.) faults; 100% pumping rate, cleaning well w 4 L/sec (#52)	Authors modelling results in USGS-Model Muse	18
29	heads & pathlines for closed (low conductivity) faults (run # 69)	Authors modelling results in USGS-Model Viewer	18
30	heads & pathlines for open (high cond.) faults (run # 52)	Authors modelling results in USGS-Model Viewer	18
31	budget for closed (low conductiv.) faults, 100% pumping rate, cleaning well w 4 L/sec (#69)	Authors modelling results in USGS-GW Chart	19
32	budget for open (high cond.) faults, 100% pumping rate,	Authors modelling results in USGS-GW Chart	19

Figure	Title	Source	page
	cleaning well w 4 L/sec (#52)		
33	Continued pollution despite functioning cleaning well for low conductivity faults (# 70)	Authors modelling results in USGS-MODPATH	20
34	Continued pollution despite functioning cleaning well for high conductivity faults (# 71)	Authors modelling results in USGS-MODPATH	20

YOUR KNOWLEDGE HAS VALUE

- We will publish your bachelor's and
  master's thesis, essays and papers

- Your own eBook and book -
  sold worldwide in all relevant shops

- Earn money with each sale

Upload your text at www.GRIN.com
and publish for free